Abdelhafid Mimouni

DNA repair: B vitamins and iron

Abdelhafid Mimouni

DNA repair: B vitamins and iron

ScienciaScripts

Cover image: www.ingimage.com

This book is a translation from the original published under ISBN 978-620-6-72015-7.

Publisher:
Sciencia Scripts
is a trademark of
Dodo Books Indian Ocean Ltd. and OmniScriptum S.R.L publishing group

120 High Road, East Finchley, London, N2 9ED, United Kingdom
Str. Armeneasca 28/1, office 1, Chisinau MD-2012, Republic of Moldova, Europe
Printed at: see last page
ISBN: 978-620-8-33553-3

DNA repair: B vitamins and iron

Author:

Dr. Abdelhafid Mimouni independent researcher specialized in bioinorganic systems chemistry. Extensive expertise in macromolecular synthesis and characterization. He obtained his PhD in chemistry from the University of Paris XII in 1997 and a Diplôme des études approfondies in bioinorganic systems from the University of Paris XI in 1993.

Summary: This book explores the complex interactions between B vitamins, iron and DNA synthesis, highlighting their crucial role in cellular metabolism. It describes in detail the biochemical functions of vitamins B12, B9, and B3, as well as iron, highlighting their impact on DNA methylation, red blood cell production and porphyrin regulation. The text examines the effects of deficiencies and excesses of these nutrients, such as pernicious anemia, neuropathies and porphyrias. By integrating a bioinorganic perspective, it reveals the underlying chemical mechanisms and molecular interactions. Chapters deal with the synergies between these nutrients, their influence on DNA repair, and modern methods for optimizing their absorption. Finally, the book proposes innovative directions for future research, such as the use of nanotechnology and the development of biosensors for precise nutrient monitoring. This work is aimed at biochemistry professionals, nutrition researchers and those interested in the molecular mechanisms of metabolism.

CONTENTS

Introduction

1. Purpose of the Book

The present book, entitled B Vitamins, Iron, and DNA Synthesis: Interactions and Implications for Health, sets out to provide an in-depth, interconnected exploration of the roles of B vitamins and iron in human health, with a particular focus on their impact on DNA synthesis and their involvement in various pathologies. This book aims to serve as a comprehensive reference for researchers, healthcare professionals and anyone interested in molecular biology and clinical nutrition.

The main aim of this book is twofold:

1. Describing Complex Interactions: The book examines how B vitamins (including B12, B9, and B3) and iron interact at the molecular level to influence DNA synthesis, red blood cell formation and other crucial biological processes. We will highlight the mechanisms by which these nutrients influence cellular function and their synergy in maintaining overall health.

2. Assessing the Clinical Implications: We will explore how imbalances or deficiencies in these nutrients can contribute to various pathologies, including Biermer's anemia, iron deficiency anemia, and other associated disorders. The book will also offer practical recommendations for managing these deficiencies and optimizing health through a balanced diet and appropriate therapeutic approaches.

The book's structure is designed to guide the reader through a progressive understanding of the essential roles of B vitamins and iron, detailing their specific functions, complex interactions, and the clinical implications of their imbalances.

2. Scientific background

B vitamins and iron are essential elements in the maintenance of human health, playing crucial roles in many fundamental biological processes.

B Vitamins: This group of vitamins is involved in various aspects of cellular metabolism. Vitamin B12, or cobalamin, is essential for DNA synthesis and the formation of red blood cells, as well as for the proper functioning of the nervous system. Folic acid (vitamin B9) is essential for DNA synthesis and repair, and plays a vital role in cell division. Vitamin B3 (niacin) is a key cofactor in metabolic reactions linked to energy production and DNA repair.

Iron: Iron is a vital mineral for the body, best known for its role in the formation of hemoglobin, the protein responsible for oxygen transport in the blood. It is also an essential component of porphyrins, crucial molecules in hemoglobin formation, and ferritin, an iron storage protein. In addition, iron plays a role in numerous enzymes involved in DNA repair processes and cellular metabolism.

Diseases and imbalances: Deficiencies or imbalances in these nutrients can lead to a range of health disorders. Biermer's anemia, a form of pernicious anemia, is due to vitamin B12 deficiency, while iron deficiency anemia is caused by iron deficiency. Insufficient folic acid is often linked to megaloblastic anemia and birth defects. Imbalances in vitamin B3 levels can lead to pellagra, a disease characterized by dermatological, digestive and neurological symptoms.

This book aims to provide a detailed understanding of the mechanisms of action of these essential nutrients, their complex interaction in cellular metabolism, and their impact on health. By reviewing current research and scientific advances in these areas, we hope to offer informed perspectives and practical recommendations for improving health management through nutrition and preventive medicine.

Part I: B vitamins and cellular metabolism

Chapter 1: Vitamin B12

1. Structure and Function

1.1 Description of Vitamin B12

Vitamin B12, or cobalamin, is a water-soluble vitamin essential for several fundamental biological functions. Its complex chemical structure and multifaceted mechanism of action make it a key element in various metabolic processes.

Chemical structure of Vitamin B12

Vitamin B12 has a characteristic chemical structure consisting of a corrin core, similar to that of porphyrins but with a cobalt atom in the center. This core is made up of four linked pyrrolic rings, forming a chelate structure around the cobalt. The central cobalt is coordinated to a methyl group (in the case of methylcobalamin) or an adenosyl group (in the case of adenosylcobalamin), defining the vitamin's specific active form.

- Corrin nucleus: Similar to porphyrins, but with a cobalt atom in the center. This structure enables vitamin B12 to interact effectively with various biochemical substrates.

- Central cobalt: The cobalt atom is essential for the formation of active cofactors that enable enzymatic interactions.

Forms of Vitamin B12 and their biological roles:

1. Methylcobalamin

Methylcobalamin is an active form of vitamin B12 found mainly in the cytoplasm of cells. It plays a crucial role in the methylation of molecules, a process essential for genetic regulation and the synthesis of compounds required by the cell.

- Nucleotide methylation: Methylcobalamin is a cofactor in the conversion of homocysteine to methionine. This process is catalyzed by the enzyme methionine synthase, where methylcobalamin supplies a methyl group to methionine. The methionine thus formed is essential for the synthesis of S-adenosylmethionine (SAMe), a universal methyl group donor in DNA and protein methylation.

- Methionine synthesis: The methyl group of methylcobalamin is transferred to homocysteine to form methionine. Methionine is then converted to SAMe, which is involved in various methylation processes, including DNA methylation. This methylation is crucial for regulating gene expression and preventing mutations.

2. Adenosylcobalamin

Adenosylcobalamin is the form of vitamin B12 found in mitochondria, the cell's energy centers. It is involved in fatty acid metabolism and energy production.

- Fatty acid metabolism: In mitochondria, adenosylcobalamin is a cofactor for the enzyme methylmalonyl-CoA mutase. This enzyme catalyzes the conversion of methylmalonyl-CoA to succinyl-CoA, a key step in the Krebs cycle. This process is crucial for the metabolism of long-chain fatty acids and the production of energy from lipids.

- Energy production: Adenosylcobalamin contributes indirectly to energy production by facilitating the steps of the Krebs cycle, a process that generates high-energy molecules such as ATP, needed for various cellular processes.

Biochemical mechanisms

- Homocysteine-Methionine Conversion: The methylation cycle, in which methylcobalamin plays a central role, is crucial for maintaining adequate levels of methionine and SAMe, required for nucleotide and protein methylation. Vitamin B12 deficiency disrupts this cycle, leading to an accumulation of homocysteine, which can have adverse effects on vascular health and genetic stability.

- Conversion of methylmalonyl-CoA to succinyl-CoA: by facilitating this conversion, adenosylcobalamin enables the products of lipid metabolism to enter the Krebs cycle, essential for energy generation. A deficiency in this form of vitamin B12 can lead to an accumulation of methylmalonyl-CoA, contributing to neurological and metabolic disorders.

In conclusion, vitamin B12 is a vital nutrient whose various forms play specific and complementary roles in cellular metabolism, DNA synthesis and energy production. Understanding these biochemical mechanisms is essential for understanding the consequences of vitamin B12 deficiency, and for developing effective treatment strategies.

1.2 Main functions of Vitamin B12

Vitamin B12 is essential for various physiological processes, profoundly affecting cellular health, red blood cell formation and nerve function. Here's a detailed explanation of its main biological functions, focusing on the underlying biochemical mechanisms:

1. DNA synthesis

Role of Vitamin B12 in DNA methylation

Vitamin B12 is a crucial cofactor in the DNA methylation process, which is essential for the regulation of gene expression and genome stability.

- Cofactor for methylation: Methylcobalamin, the active form of vitamin B12, is involved in the conversion of homocysteine to methionine. This reaction is catalyzed by the enzyme methionine synthase, where methylcobalamin provides the methyl group required for this conversion. The methionine thus formed is then converted to S-adenosylmethionine (SAMe), a methyl group donor in nucleotide methylation.

- DNA methylation: The methylation product SAMe is involved in the methylation of DNA nucleotide bases, particularly cytosines. This methylation is crucial for the regulation of gene expression, protection against mutations and genetic stability. Vitamin B12 deficiency disrupts this process, leading to DNA hypomethylation, which can result in unregulated changes in gene expression and genetic abnormalities.

2. Red blood cell formation

Importance of Vitamin B12 for Erythropoiesis

Vitamin B12 plays an essential role in the production and maturation of red blood cells, directly influencing erythropoiesis, the process by which red blood cells are formed in the bone marrow.

- Red blood cell production: Vitamin B12 is required for DNA synthesis in blood cells. Its deficiency leads to abnormal production of red blood cells, notably inefficient erythropoiesis, resulting in large, immature red blood cells known as megaloblasts.

- Prevention of Megaloblastic Anemia: Defective red blood cells produced in the absence of vitamin B12 cannot carry oxygen efficiently, resulting in anemia symptoms such as fatigue, pallor and weakness. Vitamin B12 deficiency is therefore directly linked to megaloblastic anemia, a condition where red blood cells are abnormally large and poorly functioning.

3. Nerve function

Role of Vitamin B12 in Nervous System Health

Vitamin B12 is essential for maintaining the health of the central nervous system, notably by helping to form and maintain myelin, a lipid substance that surrounds and protects nerve fibers.

- **Myelin formation**: Adenosylcobalamin, an active form of vitamin B12 found in mitochondria, is essential for the biosynthesis of myelin lipids. It plays a role in the synthesis of the essential fatty acids that make up myelin. This protective membrane is crucial for the efficient conduction of nerve impulses along axons.

- **Nerve function and neuropathies**: Vitamin B12 deficiency leads to demyelination of the nerves, which can cause peripheral neuropathies and neurological disorders such as tingling, numbness, coordination disorders and cognitive deficits. Demyelination is also associated with more serious neurological disorders, such as motor disorders and sensory deficits.

In summary, vitamin B12 is essential for DNA synthesis, the formation of red blood cells, and the maintenance of nerve health. Its deficiency can cause major disruptions in these processes, leading to clinically significant conditions such as megaloblastic anemia and neurological disorders. Understanding these biochemical mechanisms is crucial to the effective diagnosis and treatment of vitamin B12 deficiency.

2. Role in DNA synthesis

2.1 Mechanism of action in DNA methylation

Vitamin B12 is essential for DNA methylation, a process crucial to the regulation of gene expression and genome stability. DNA methylation plays a central role in transcriptional regulation, chromatin organization and DNA repair. Here's a detailed look at the biochemical mechanisms by which vitamin B12 influences this process::

1. Role of Methylcobalamin in the Methylation Cycle

Methylcobalamin, an active form of vitamin B12, is directly involved in the conversion of homocysteine to methionine, a process catalyzed by the enzyme methioninesynthase. This process is essential for the production of S-adenosylmethionine (SAMe), a methyl group donor.

- Conversion of Homocysteine to Methionine: Methylcobalamin provides a methyl group to homocysteine to form methionine. This reaction is crucial for maintaining adequate levels of methionine, which is then converted to SAMe.

- Formation of S-adenosylmethionine (SAMe): Methionine, obtained via the methylation cycle, is converted to SAMe, which plays a key role as a methyl group donor in various biochemical reactions, including DNA methylation.

2. DNA methylation

SAMe is used in the methylation of DNA nucleotides, particularly cytosines. This process is crucial for many aspects of gene regulation and genome stability.

- Cytosine methylation: DNA methylation involves the addition of a methyl group (CH_3) to the 5-position of cytosines in gene promoter regions. This methylation can modulate gene expression by affecting the accessibility of DNA to transcriptional machinery. Excessive or insufficient methylation can lead to changes in gene expression, contributing to diseases such as cancer.

- Regulation of Gene Expression: Methylated regions of DNA are often associated with reduced gene expression, by preventing transcription factors and enzymes required for gene activation from accessing the DNA. Thus, methylation is a key mechanism for epigenetics, enabling the control of gene expression without altering the DNA sequence itself.

- Genome stability: DNA methylation also contributes to genome stability by preventing chromosomal rearrangements and promoting DNA damage repair. Inadequate methylation can lead to genetic instabilities, increasing the risk of mutations and genetic diseases.

3. Implications of Vitamin B12 deficiency

Vitamin B12 deficiency compromises the methylation cycle, leading to reduced SAMe levels and altered DNA methylation. The consequences of this disruption are manifold:

- **Hypomethylation**: Vitamin B12 deficiency can lead to hypomethylation of cytosines in gene promoter regions, which can result in aberrant gene expression and increased susceptibility to diseases, including certain types of cancer.

- **Alteration of genome stability**: disruption of DNA methylation can compromise genome stability, increasing the risk of mutations and chromosomal rearrangements. This can contribute to genetic pathologies and disorders linked to genetic instability.

In summary, vitamin B12, via methylcobalamin, plays a central role in DNA methylation as a cofactor for SAMe formation. This process is crucial for the regulation of gene expression, genomic stability, and the prevention of diseases linked to aberrant methylation. Vitamin B12 deficiency can have a significant impact on these fundamental biological processes.

2.2 Interaction with the Methylation Cycle and Importance for DNA Repair

The methylation cycle is a complex biochemical network in which vitamin B12 plays a crucial role. This cycle influences various aspects of gene regulation, DNA repair and genomic stability. Here's a detailed explanation of these interactions:

1. Interaction with the Methylation Cycle

The methylation cycle is a metabolic process in which methyl groups (CH_3) are transferred between different molecules, thus influencing various biological processes, including DNA methylation. Vitamin B12 is mainly involved through methylcobalamin, which is essential for maintaining a proper balance of methyl group donors.

1.1 Conversion of homocysteine to methionine

- Methionine Synthase cofactor: Methylcobalamin, the active form of vitamin B12, is a cofactor for the enzyme methionine synthase, which catalyzes the conversion of homocysteine to methionine. This process is crucial for the formation of S-adenosylmethionine (SAMe), the main methyl group donor in the methylation cycle.

- SAMe formation: Methionine is converted to SAMe, which provides methyl groups for various methylation processes, including DNA methylation. Adequate availability of SAMe depends on the efficiency of vitamin B12 in this process.

1.2 Importance of SAMe in the Methylation Cycle

- Methyl group donor: SAMe transfers methyl groups to various substrates, including DNA, proteins and lipids. This transfer is essential for the regulation of gene expression, myelin formation and other biological processes.

- Regulation of the methylation cycle: Vitamin B12 deficiency can lead to a reduction in SAMe levels, disrupting the methylation cycle and affecting various essential biological processes.

2. Importance for DNA repair

DNA methylation, influenced by vitamin B12, plays a key role in DNA damage repair and genomic stability.

2.1. DNA repair

- Damage repair: DNA methylation is involved in DNA repair mechanisms, notably base excision repair. This process is crucial for correcting errors caused by environmental damage or replication errors.

- Genomic stability: Proper DNA methylation ensures genome stability by preventing chromosomal rearrangements and maintaining the integrity of genetic information. Disruption of this process, due to vitamin B12 deficiency, can lead to mutations and genetic abnormalities.

2.2. Regulation of gene expression

- DNA accessibility: Methylation of cytosines in DNA promoter regions modifies chromatin conformation, making DNA more or less accessible to transcriptional machinery. Excessive or insufficient methylation can influence gene expression by inhibiting or activating transcription.

- Expression control: For example, high levels of methylation can lead to gene repression, while low levels can result in aberrant expression. Vitamin B12, by influencing the methylation cycle, therefore plays a role in the fine regulation of gene expression.

2.3 Implications of Vitamin B12 deficiency

- Hypomethylation and genomic instability: Vitamin B12 deficiency can result in DNA hypomethylation, leading to increased genomic instability and susceptibility to mutations. This instability is often associated with diseases such as cancer and neurological disorders.

- Disruption of DNA repair: Reduced SAMe levels can affect DNA repair mechanisms, increasing the risk of unrepaired damage and pathological genetic changes.

In conclusion, vitamin B12 is essential for the methylation cycle, directly influencing DNA methylation, DNA repair and the regulation of gene expression. Its deficiency can have significant consequences on genomic stability and cellular health, increasing the risk of various diseases.

3. Deficiency and associated diseases

3.1 Pernicious anemia (Biermer's anemia)

Pernicious anemia, also known as Biermer's anemia, is a specific form of anemia due to vitamin B12 deficiency. This condition is often caused by a deficiency in intrinsic factor, a protein essential for the absorption of vitamin B12 in the intestine. Here's a detailed overview of this disease and its manifestations:

1. Etiological mechanism

- **Intrinsic factor deficiency:** Pernicious anemia generally results from the absence or reduced production of intrinsic factor, a glycoprotein secreted by the parietal cells of the stomach. This protein is crucial for vitamin B12 binding and absorption in the ileum, a part of the small intestine. Without intrinsic factor, vitamin B12 cannot be absorbed efficiently.

- **Auto-immune origin**: Intrinsic factor deficiency is often caused by an auto-immune reaction, where the immune system attacks the stomach's parietal cells, reducing production of this essential protein. This condition leads to inadequate absorption of vitamin B12, resulting in insufficient blood levels.

2. Clinical symptoms

- **Fatigue and weakness:** Vitamin B12 deficiency causes insufficient production of red blood cells, resulting in megaloblastic anemia. Red blood cells are larger than normal and less efficient at carrying oxygen, leading to symptoms of persistent fatigue, generalized weakness and pallor.

- **Neurological dysfunction:** The neurological effects of vitamin B12 deficiency are diverse and may include:

o **Paresthesias**: tingling or numbness in the hands and feet, due to damage to peripheral nerves.

o **Walking disorders**: Difficulties in coordination and stability during walking, often caused by damage to the nerve pathways responsible for proprioception and motor control.

o **Cognitive disorders**: memory and concentration problems, as well as difficulty in performing complex cognitive tasks, due to neurotoxic effects on the central nervous system.

3.2 Neuropathies and other clinical manifestations

Vitamin B12 deficiency can also lead to a variety of other neurological and clinical manifestations, in addition to the symptoms of pernicious anemia:

1. Peripheral neuropathies

- **Nerve pain**: Peripheral neuropathies due to vitamin B12 deficiency can cause acute or chronic nerve pain, resulting from demyelination of nerve axons. These pains can range from burning to stabbing, and can affect patients' quality of life.

- **Coordination disorders**: Demyelination of sensory and motor nerves can lead to coordination and balance disorders. Patients may have difficulty with fine movements and complex motor tasks.

2. Gastrointestinal disorders

- **Digestive symptoms**: Vitamin B12 deficiency can also lead to gastrointestinal symptoms such as nausea, vomiting, abdominal pain, and changes in bowel habits, often due to impaired digestive function and altered nutrient absorption.

In conclusion, vitamin B12 deficiency can cause significant and varied symptoms, including pernicious anemia, peripheral neuropathies, gastrointestinal disorders, and neurological dysfunction. A thorough understanding of these manifestations is essential for early diagnosis and effective management of vitamin B12 deficiency.

Chapter 2: Folic acid (vitamin B9)

1. Structure and Function

Folic acid, or vitamin B9, is a water-soluble vitamin essential to a variety of biological processes. Its chemical structure is composed of a pyridine nucleus, para-aminobenzoic acid (PABA), and one or more glutamate residues. In the human body, folic acid exists in several biologically active forms, the main ones being:

- **Tetrahydrofolate (THF)**: The active form of folic acid, resulting from the reduction of folic acid in cells. THF plays a crucial role in the transport and transferability of the monocarbon groups required for nucleotide and amino acid synthesis.

- **Methyltetrahydrofolate (5-MTHF)**: The main form of folic acid in blood plasma, involved in the methylation of homocysteine to methionine, essential for various biochemical reactions, including DNA methylation.

Folic acid is essential for cell growth and the formation of red blood cells. It is involved in the synthesis of nucleic acids and amino acids, and helps maintain cellular health and prevent neural tube defects.

2. Role in DNA synthesis

Folic acid plays a central role in DNA synthesis through several biochemical mechanisms:

- **Nucleotide synthesis**: Folic acid is essential for the synthesis of nucleotides, the building blocks of DNA. 5-MTHF provides the carbon units needed to synthesize thymidine, one of DNA's four nitrogenous bases. This synthesis is crucial for DNA replication and repair.

- **DNA repair**: Folic acid is involved in DNA repair, providing the carbon units needed for methylation and repair processes. Folic acid deficiency can compromise DNA damage repair, increasing the risk of genetic mutations and cancers.

- **Folate cycle**: Folic acid is involved in the folate cycle, where it is converted to THF. This cycle is crucial for providing monocarbon groups for nucleotide synthesis and for the metabolism of homocysteine, an amino acid whose excess can be toxic and is associated with cardiovascular disease.

3. Deficiency and associated diseases

Folic acid deficiency can have serious health consequences:

- **Megaloblastic anemia**: This form of anemia is characterized by the presence of large, immature red blood cells in the blood. Folic acid deficiency disrupts DNA synthesis, leading to abnormal production and reduced numbers of red blood cells.

- **Risks for Pregnancy and Congenital Malformations**: Folic acid deficiency during pregnancy is associated with an increased risk of congenital malformations, such as neural tube defects (e.g. spina bifida). Adequate folic acid supplementation before and during the first weeks of pregnancy is crucial for normal fetal development.

Chapter 3: Vitamin B3 (Niacin)

1. Structure and Function

Vitamin B3, also known as niacin, is a water-soluble vitamin essential for various metabolic functions in the human body. It comes in two main biological forms:

- **Nicotinic acid (or Niacin)**: This form of vitamin B3 is best known for its effect on blood lipids. Nicotinic acid is used to reduce LDL (low-density lipoprotein) and triglyceride levels in the blood, while increasing HDL (high-density lipoprotein) levels. This action contributes to the prevention of cardiovascular disease.

- **Nicotinamide (or Niacinamide)**: This is the form of vitamin B3 found mainly in body tissue. Nicotinamide is converted into two essential coenzymes: NAD+ (nicotinamide adenine dinucleotide) and NADP+ (nicotinamide adenine dinucleotide phosphate). These coenzymes are crucial for the majority of redox reactions in cellular metabolism.

Metabolic functions:

- **Energy production**: NAD+ and NADP+ are essential cofactors in the respiratory chain, the Krebs cycle and glycolysis. NAD+ is crucial for the dehydrogenation of metabolic substrates, enabling the production of ATP, the main energy source for cells.

- **Fatty acid and cholesterol synthesis**: NADP+ is involved in the biosynthetic pathways of fatty acids and cholesterol. It plays a role in regulating lipid levels and in the production of cholesterol precursors, thus influencing cardiovascular health.

2. Role in DNA repair

Vitamin B3, in the form of NAD+, is directly involved in DNA repair mechanisms, as well as in the regulation of genetic transcription. Here's how it works:

- **DNA repair**: NAD+ is an essential cofactor for poly(ADP-ribose) polymerase (PARP) enzymes. When a DNA double-strand break occurs, PARPs use NAD+ to catalyze the addition of poly(ADP-ribose) groups to DNA repair proteins. This process activates the proteins needed to repair damage and restore genome integrity. NAD+ depletion can compromise the ability of PARPs to repair DNA, increasing the risk of mutations and associated diseases.

- **Transcription regulation:** NAD+ influences gene transcription by acting as a substrate for sirtuins, a family of deacetylase enzymes. Sirtuins modify chromatin structure by deacetylating histones, thereby altering DNA accessibility and regulating gene expression. This mechanism is crucial for controlling the cellular stress response and regulating DNA repair processes.

3. Deficiency and associated diseases

Vitamin B3 deficiency can lead to a number of significant health problems:

- **Pellagra**: Pellagra is a disease caused by severe niacin deficiency, and is characterized by a symptomatic triad: dermatitis, diarrhea and dementia. Dermatitis is usually asymmetrical and localized to sun-exposed areas, while diarrhea can be severe and dementia can extend to severe cognitive impairment. The disease is often associated with diets low in niacin, and can have a significant impact on quality of life.

- **Effects on Energy Metabolism**: Niacin deficiency leads to a decrease in NAD+ and NADP+ levels, which disrupts the metabolic pathways essential for energy production. This disruption can affect various biological processes, including the breakdown of carbohydrates, lipids and proteins. Symptoms can include extreme fatigue, muscle weakness, and more generalized metabolic disorders.

Part II: Iron and Porphyrins

Chapter 4: Iron and its metabolism

1. Absorption and transport

Iron absorption mechanisms

Iron is an essential element for many biological processes, and its absorption is regulated in a sophisticated way to maintain iron balance in the body. Iron absorption occurs mainly in the duodenum and proximal jejunum of the small intestine, and involves several complex mechanisms:

- **Forms of iron**: Dietary iron comes in two forms: heme iron and non-heme iron. Haem iron, found in animal products, is generally better absorbed than non-haem iron, which is found in plants. Haem iron is absorbed directly by the cells of the intestinal epithelium via a specific transporter, the haem iron transporter (HCP1). Non-heme iron requires chemical reduction prior to absorption.

- **Reduction and transport**: Non-heme iron is first reduced from ferric iron (Fe^{3+}) to ferrous iron (Fe^{2+}) by the enzyme ferrireductase, present on the surface of intestinal epithelial cells. Ferrous iron is then transported inside intestinal cells by the divalent metal transporter (DMT1).

- **Intracellular storage and transport:** Once iron has been absorbed by intestinal epithelial cells, it can be stored in the form of ferritin, a protein that preserves iron for future use. Alternatively, iron can be transported out of cells by ferroportin, a specific protein found on the cell surface. Before iron can leave the cell and enter the bloodstream, it must be oxidized from its ferrous form (Fe^{2+}) to ferric iron (Fe^{3+}). This oxidation is carried out by an enzyme called ferroxidase. Once in ferric form, iron can bind to transferrin, another protein present in blood plasma, which transports iron to the tissues and organs that need it.

Iron transport in the body

Iron transported in plasma is mainly bound to transferrin, a glycoprotein that delivers iron to tissues and organs. Transferrin binds to specific receptors on cells, enabling endocytosis of the iron-transferrin complex. Inside cells, iron is released and used for various biological processes, or stored as ferritin.

Iron transport and regulation are orchestrated by regulatory proteins such as hepcidin, a hormone produced by the liver in response to body iron levels. Hepcidin inhibits the release of iron from the intestine and iron stores, thus regulating iron absorption and storage according to the body's needs.

2. Physiological roles

Involvement in Hemoglobin Formation and Porphyrin Synthesis

- Hemoglobin formation: Iron is an essential component of hemoglobin, the red blood cell protein responsible for transporting oxygen from the lungs to the tissues and carbon dioxide from the tissues to the lungs. Each hemoglobin contains four heme groups, each composed of a porphyrin nucleus with an iron atom in the center. Iron is crucial to hemoglobin's ability to bind oxygen, and its absence or deficiency leads to disturbances in tissue oxygenation, as in iron-deficiency anemia.

- Porphyrin synthesis: Porphyrins are cyclic structures that bind to iron to form complexes such as hemoglobin and myoglobin. Iron plays a crucial role in porphyrin synthesis, being necessary for the formation of the porphyrin core. Porphyrin synthesis begins in the mitochondria, where glycine and succinyl-CoA condense to form delta-aminolevulinic acid (ALA). ALA is then converted to porphobilinogen and finally to protoporphyrin, which combines with iron to form heme.

Role in Enzymes and Energy Metabolism

Iron is also an essential cofactor for numerous enzymes involved in various metabolic processes:

- **Oxidoreductase enzymes**: Iron is a cofactor for several oxidoreductase enzymes, such as cytochromes P450, which are involved in lipid, steroid and drug metabolism. These enzymes play a key role in the detoxification of foreign substances and the synthesis of bioactive molecules.

- **Energy metabolism**: Iron is essential to the electron transport chain in mitochondria, where it is a component of complexes I, II, III and IV. These complexes play a crucial role in the production of ATP, the main source of cellular energy. Iron is also involved in neurotransmitter synthesis and amino acid metabolism.

3. Deficiency and associated diseases

Iron deficiency anemia

Iron deficiency anemia is the most common blood disorder caused by iron deficiency. It is characterized by a decrease in the levels of iron available for hemoglobin synthesis, leading to insufficient production of red blood cells and impaired function:

- Symptoms: Symptoms include fatigue, weakness, pallor, shortness of breath, and sometimes heart palpitations. In severe cases, iron-deficiency anemia can lead to cognitive impairment and cardiac complications.

- Diagnosis and treatment: Diagnosis is made by blood tests measuring levels of ferritin, serum iron, and total iron-binding capacity (TIBC). Treatment usually involves oral iron supplementation or, in some cases, iron injections, as well as evaluation of the underlying causes of deficiency.

Effects of Iron Overdosage and Related Diseases (Hemochromatosis)

- Hemochromatosis: Hemochromatosis is a genetic or secondary disease characterized by excessive iron absorption and abnormal iron deposition in various tissues, including the liver, heart and pancreas. This iron overload can lead to serious complications, such as liver disease (hepatitis, cirrhosis), endocrine disorders (diabetes, thyroid insufficiency), and cardiac disorders (cardiomyopathy).

- Symptoms: Symptoms of hemochromatosis include chronic fatigue, joint pain, gray or bronze skin, and metabolic disorders. Early diagnosis is essential to prevent serious complications. Treatments include regular bloodletting to reduce iron levels and dietary modifications.

- Acute iron overdose: An acute excess of iron, often due to an overdose of iron supplements, can cause acute symptoms of toxicity such as abdominal pain, vomiting and gastrointestinal disorders. In severe cases, acute overload can lead to organ damage and potentially fatal risks.

Chapter 5: Porphyrins

1. Porphyrin structure and function

Porphyrins description

Porphyrins are complex cyclic structures composed of four pyrrolic units linked by methine bridges, forming a tetra-pyrrolic core. This core is the foundation of several crucial biological molecules, including heme, chlorophyll and cobalamin (vitamin B12). The basic structure of porphyrins is an alternation of double bonds, giving these molecules their electronic properties and their ability to bind metal ions.

- Heme: The best-known porphyrin is heme, in which the porphyrin nucleus binds to an iron atom (Fe^{2+}), forming a complex called ferroprotoporphyrin. Heme is an essential component of hemoglobin and myoglobin, the proteins responsible for transporting and storing oxygen in cells.

- Chlorophyll: In plants, a similar porphyrin, chlorophyll, contains a magnesium atom (Mg^{2+}) in the center of its nucleus, playing a crucial role in photosynthesis.

- Cobalamin: Vitamin B12 also has a modified porphyrin structure, in which the corrin nucleus surrounds a cobalt atom (Co^{2+}), essential for several enzymatic processes.

Role in Hemoglobin and Ferritin Formation

- Hemoglobin formation: Hemoglobin is composed of four protein subunits, each containing a heme group. The iron in heme is responsible for reversible oxygen binding, enabling oxygen transport from the lungs to the tissues. Hemoglobin synthesis is regulated by the availability of iron and porphyrins. Heme groups are formed in the mitochondria of erythroid cells, where iron is incorporated into the porphyrin core to create heme.

- Ferritin formation: Ferritin is an iron storage protein, composed of a core of oxidized iron stored as ferrihydrite, surrounded by a protein shell. Ferritin synthesis is influenced by available iron levels. Increased iron levels in the body stimulate ferritin production to store excess iron, while iron deficiency inhibits this production.

2. Interconnection with the Iron

Porphyrin synthesis and regulation

Porphyrin synthesis and regulation are closely linked to iron levels in the body. Iron is a crucial component of the heme group, and its metabolism directly influences porphyrin synthesis.

- Porphyrin synthesis: Heme synthesis begins in the mitochondria with the condensation of glycine and succinyl-CoA to form delta-aminolevulinic acid (ALA). This reaction is catalyzed by 5-aminolevulinate synthase (ALAS), an enzyme regulated by iron levels. ALA is then converted to porphobilinogen, then to protoporphyrin IX, which combines with iron to form heme.

- Iron regulation: Iron levels influence the regulation of enzymes involved in porphyrin synthesis. Sufficient iron stimulates heme production, while insufficient iron levels reduce the activity of ALAS and other key enzymes, thus limiting heme production.

- Inhibitory feedback: heme itself exerts a negative feedback effect on porphyrin synthesis. High levels of heme inhibit ALAS production, thus regulating the amount of heme produced according to cellular needs.

Interactions with Iron Metabolism

The metabolism of iron and porphyrins is also influenced by iron transport and storage mechanisms. Iron absorption, distribution and storage in the form of ferritin affect the availability of iron for heme synthesis.

- Iron transport and storage: Transferrin transports iron in plasma to cells, where it is either incorporated into porphyrins to form heme, or stored in ferritin. Transferrin and ferritin levels are regulated according to iron requirements and iron saturation levels in the body.

- Iron balance: An imbalance in iron levels can lead to abnormalities in porphyrin synthesis. Iron overload can lead to excessive hemoglobin accumulation and disorders such as hemochromatosis, while iron deficiency can result in reduced hemoglobin and porphyrin production.

3. Deficiency and associated diseases

Iron deficiency anemia

- Pathophysiology: Iron deficiency anemia is caused by iron deficiency, resulting in reduced red blood cell production and impaired hemoglobin formation. This results in reduced red blood cell size (microcytosis) and a lower-than-normal hemoglobin concentration.

- Symptoms: Symptoms include fatigue, weakness, pallor, dizziness and shortness of breath. In the long term, severe iron deficiency can lead to cardiac complications and cognitive impairment.

- Diagnosis and treatment: Diagnosis is based on blood tests measuring levels of ferritin, serum iron and total iron-binding capacity (TIBC). Treatment usually consists of iron supplementation and investigation of the underlying causes of deficiency.

Porphyrias

Porphyrias are a group of rare metabolic disorders caused by abnormalities in the synthesis of porphyrins and heme. They can be hereditary or acquired, and present a variety of symptoms depending on the type of porphyria.

- Acute Intermittent Porphyria: This form of porphyria is caused by porphobilinogen deaminase deficiency, leading to acute attacks of abdominal pain, nausea and neurological disturbances. Seizures are often triggered by factors such as medications, infections, or prolonged fasting.

- Porphyria Cutanea Tardive: Caused by uroporphyrinogen decarboxylase deficiency, this form of porphyria manifests as photosensitivity, skin lesions and discoloration. Symptoms are exacerbated by exposure to the sun, and can lead to scarring and skin deformities.

- Erythropoietic porphyria: This rare form is due to a deficiency in ferrochelatase, resulting in an accumulation of heme in erythrocytes. Symptoms include anemia, abdominal pain and abnormal skin pigmentation.

- Diagnosis and treatment: Diagnosis of porphyrias relies on biological tests to measure porphyrin levels in urine, stool and plasma. Treatment varies according to the type of porphyria, and may include dietary modifications, medication to alleviate symptoms, and in some cases, surgery to manage complications.

Part III: Interactions between B vitamins, iron and DNA synthesis

Chapter 6: Biological interactions

1. Nutrient synergy

B vitamins and iron interact in a complex and synergistic way in various biological processes essential to human health. Here's a detailed analysis of these interactions:

1.1 B Vitamins and Iron: A Metabolic Relationship

B vitamins, in particular B9 (folic acid), B12 (cobalamin), and B6 (pyridoxine), as well as iron, play crucial roles in cellular metabolism and blood cell production. These nutrients work together to ensure optimal functioning of metabolic and biological processes.

- **Folic acid and iron**: Folic acid (vitamin B9) and iron are essential for the formation of red blood cells. Folic acid is involved in the synthesis of nucleotides, the building blocks of DNA, which is crucial for the production of red blood cells in the bone marrow. Iron, meanwhile, is a central component of hemoglobin, the protein responsible for transporting oxygen in the blood. Folic acid deficiency can lead to megaloblastic anemia, where red blood cells are abnormally large and immature, a condition that can also be exacerbated by iron deficiency.

- **Vitamin B12 and Iron**: Vitamin B12 (cobalamin) is essential for the synthesis of methionine from homocysteine, a process which also produces S-adenosylmethionine (SAMe), a methyl group donor in various biochemical reactions. Vitamin B12 deficiency disrupts this pathway, leading to an accumulation of homocysteine and a failure in red blood cell production, which can also lead to iron deficiency due to impaired absorption and inefficient erythropoiesis.

- **Vitamin B6 and iron**: Vitamin B6 (pyridoxine) is involved in amino acid metabolism, including the conversion of homocysteine to cysteine. Vitamin B6 deficiency can lead to high levels of homocysteine, disrupting hemoglobin synthesis and contributing to iron-deficiency anemia.

1.2 Common metabolic pathways

B vitamins and iron share several metabolic pathways and can influence each other's efficacy:

- **Hemoglobin synthesis**: Vitamins B9 and B12 are necessary for the synthesis of hemoglobin, the oxygen-carrying protein. A deficiency in folic acid or vitamin B12 disrupts this synthesis, leading to anemia, even if sufficient iron reserves are available.

- **Methionine metabolism**: Vitamin B12 is essential for the methylation of homocysteine to methionine. This reaction is dependent on the presence of folic acid to provide the necessary methyl groups. Iron is also involved in this metabolic pathway via specific enzymes that require iron as a cofactor.

- **DNA repair**: Vitamin B9 and B12 influence DNA methylation, a process crucial to DNA repair. Iron is necessary for the function of certain enzymes involved in DNA repair, such as ribonucleotide reductases.

2. Impact on DNA synthesis

B vitamins and iron play complementary roles in DNA synthesis and repair, each making specific contributions to cellular processes.

2.1 Role of B vitamins

- **Folic acid (B9)**: Folic acid is essential for the synthesis of nucleotides, the basic units of DNA. It provides the methyl groups needed to form purine and pyrimidine bases. Folic acid deficiency leads to altered DNA synthesis and an increased risk of genetic mutations. Folic acid also influences DNA methylation, a process crucial to the regulation of gene expression and genome stability.

- **Vitamin B12:** Vitamin B12 is essential for the conversion of homocysteine to methionine, a reaction which also produces SAMe, a methyl group donor for DNA methylation. Vitamin B12 deficiency disrupts DNA methylation, affecting DNA repair and gene expression. This can lead to genetic abnormalities and cell proliferation disorders.

2.2 Iron's role

- **DNA synthesis**: Iron is a crucial cofactor for several enzymes involved in DNA synthesis. It is required for the function of ribonucleotide reductase, which converts ribonucleotides into deoxyribonucleotides, the building blocks of DNA. Iron deficiency leads to a reduction in the production of these deoxyribonucleotides, disrupting DNA synthesis and slowing cell proliferation.

- **DNA repair**: Iron is also involved in DNA repair mechanisms via proteins such as DNA glycosylases and DNA polymerases, which require iron for their enzymatic activity. Iron contributes to the repair of oxidative DNA damage, which is crucial for maintaining genomic integrity.

Chapter 7: Case studies and associated anomalies

1. Biermer anemia

Biermer anemia is a specific condition linked to vitamin B12 deficiency, often due to a deficiency in intrinsic factor, a protein crucial for the absorption of vitamin B12 in the digestive system. This chapter explores in detail the mechanisms underlying this deficiency and its effects on the body.

1.1 The role of vitamin B12

Vitamin B12, also known as cobalamin, is essential for a number of biological functions, including the formation of red blood cells and maintenance of the nervous system. It plays a key role in the conversion of homocysteine to methionine, a process crucial to the production of S-adenosylmethionine (SAMe), which is involved in the methylation of molecules and the regulation of gene expression.

1.2 Pathological mechanisms

Vitamin B12 deficiency due to a lack of intrinsic factor leads to a disturbance in the vitamin's absorption. This lack affects DNA synthesis and red blood cell formation, leading to insufficient production of these cells and an increase in their size, characteristic of Biermer anemia. This condition is also associated with neurological abnormalities, such as impaired coordination and abnormal sensations.

2. Other associated anomalies

2.1 Megaloblastic anemia

Megaloblastic anemia is a form of anemia resulting from deficiencies in folic acid or vitamin B12, which cause a disturbance in the formation of red blood cells. This deficiency leads to the production of large, immature red blood cells, compromising their oxygen-carrying function. Abnormalities in DNA synthesis due to a lack of these vitamins are at the heart of this type of anemia.

2.2 Iron deficiency anaemia

Iron deficiency anemia is caused by a lack of iron, which is needed to form hemoglobin, the protein in red blood cells that carries oxygen. Insufficient iron intake or impaired absorption leads to a reduced number of red blood cells and symptoms such as fatigue and paleness. This condition is often the result of nutritional deficiency or blood loss.

2.3 Other associated conditions

Other vitamin and iron imbalances can also cause biological abnormalities. For example, vitamin B6 deficiency can disrupt amino acid metabolism and iron absorption, while excess iron can lead to conditions such as hemochromatosis, a disease characterized by iron overload in body tissues.

Chapter 8: Diagnosis and management of vitamin and iron deficiencies

This chapter discusses how chemical and bioinorganic techniques are applied to identify and treat B-vitamin and iron imbalances. Emphasis is placed on analytical methods and bioinorganic strategies for understanding and correcting these deficiencies.

1. Diagnosis of deficiencies

1.1 Diagnostic tests

Diagnostic tests for B-vitamin and iron deficiencies are essential to accurately assess the levels of these nutrients in the body. Here are the advantages of each method:

Atomic Absorption Spectroscopy (AAS)

- **High precision**: SAA enables precise measurement of iron concentrations by detecting the absorption of light by excited iron atoms. This precision is essential for determining exact iron levels in biological samples.
- **Sensitivity**: This method is highly sensitive, capable of detecting very low concentrations of iron, which is crucial for diagnosing iron deficiency at an early stage.
- **Speed**: Results can be obtained quickly, which is important for rapid diagnosis and appropriate intervention.
- **Sample simplicity**: SAA often requires minimal sample preparation, which simplifies the testing process.

High-Performance Liquid Chromatography (HPLC)

- **Precise separation**: HPLC separates the various B vitamins in a sample according to their interactions with the stationary and mobile phase, offering precise separation of compounds.

- **Reliable quantification**: This method provides reliable quantifications of B vitamins, helping to assess the levels of each vitamin in detail.
- **Versatility**: HPLC can be used to analyze a wide range of biological compounds, including vitamins, with high resolution.
- **Interference reduction**: Efficient separation of components minimizes interference from other substances in the sample.

Enzyme-linked immunosorbent assay (ELISA)

- **High specificity**: ELISA uses specific antibodies to target and quantify B vitamins, offering highly specific detection for each vitamin.
- **Sensitivity**: This technique is extremely sensitive and can detect very low concentrations of vitamins, which is essential for diagnosing deficiencies at an early stage.
- **Ease of use**: ELISA tests are relatively simple to perform and interpret, making them accessible for use in a variety of laboratories.
- **Fast quantification**: ELISA provides fast, accurate results, enabling efficient assessment of vitamin levels.

Mass Spectrometry

- **Detailed analysis**: Mass spectrometry provides a detailed analysis of the molecules present in biological samples, including B vitamins and iron-related compounds, enabling an in-depth understanding of nutrient levels.
- **Metabolite Identification**: This is particularly useful for identifying specific metabolites and assessing nutrient levels with great precision, providing information on vitamin and mineral metabolism.

- **Versatility**: This technique can analyze a wide variety of compounds in a single sample, making it possible to detect several nutritional deficiencies or imbalances simultaneously.
- **High accuracy**: Mass spectrometry is known for its high accuracy and ability to provide reliable quantitative data on nutrient concentrations in samples.

These diagnostic techniques all play an important role in assessing B-vitamin and iron levels, each offering unique advantages that enable accurate and reliable analysis of nutritional deficiencies.

1.2 Biochemical detection methods

Biochemical methods play a crucial role in the diagnosis of deficiencies. For example:

- Methylmalonyl-CoA and Homocysteine assays: High levels of methylmalonyl-CoA and homocysteine in plasma may indicate vitamin B12 deficiency, as these molecules accumulate when vitamin B12 is insufficient to metabolize them.

- Ferritin Marker Assessment: Ferritin, an iron storage protein, is a key indicator of the body's iron stores. Decreased ferritin levels may signal iron deficiency, while elevated levels may indicate iron overload.

2. Processing and management

2.1 Therapeutic approaches

Strategies for correcting B-vitamin and iron deficiencies are based on bioinorganic and chemical principles. Here are some key approaches:

- B vitamins and iron supplementation: Supplements are used to restore adequate levels of these nutrients. For example, iron salts (such as ferrous sulfate) are often prescribed to treat iron deficiency, while vitamin B12 and folate preparations are administered to correct deficiencies in these vitamins.

- Supplement formulation: Food supplements can be formulated to improve nutrient absorption. For example, chelated forms of iron, which are better absorbed by the body, are used to treat deficiencies more effectively.

- Diet modification: A practical approach involves fortifying foods with B vitamins and iron. For example, cereal products can be fortified with B vitamins and iron to improve overall nutritional intake.

2.2 Optimizing Absorptions

- Nutrient interactions: Interactions between B vitamins, iron and other nutrients can influence treatment efficacy. For example, vitamin C enhances iron absorption, which can be used to optimize the correction of iron deficiency.

- Use of Active Forms: Biologically active forms of B vitamins, such as methylcobalamin and 5-methyltetrahydrofolate, are often preferred for their greater availability and effectiveness in treating deficiencies.

Chapter 9: Nutritional recommendations

This chapter examines nutritional recommendations for B vitamins and iron, focusing on recommended daily intakes and dietary approaches to maintaining an optimal balance of these essential nutrients. We'll explore guidelines established by health organizations, strategies for integrating these nutrients into daily diets, and the impact of these practices on overall health.

1. Recommended intakes

1.1 General recommendations

Recommendations for daily intake of B vitamins and iron are established to ensure sufficient absorption of these nutrients to support normal metabolic and physiological functions. These recommendations vary according to age, gender and specific needs, such as those of pregnant or breast-feeding women.

- Vitamin B1 (Thiamine): Recommended intake is generally 1.1 to 1.2 mg per day for adults. Requirements may be higher in people with rapid metabolism or poor diets.

- Vitamin B2 (Riboflavin): Recommended at 1.1 mg for women and 1.3 mg for men per day. This vitamin is crucial for redox reactions and energy production.

- Vitamin B6 (Pyridoxine): Recommended daily intake is 1.3 to 2.0 mg, depending on age and sex. It plays a role in amino acid metabolism and neurotransmitter production.

- Vitamin B12 (Cobalamin): Adults need 2.4 µg per day. This vitamin is essential for the formation of red blood cells and the maintenance of neurological health.

- Folate (folic acid): The recommended intake is 400 µg per day. The requirement increases to 600 µg for pregnant women, due to its crucial role in DNA synthesis and prevention of congenital malformations.

- Iron: Iron requirements are 18 mg per day for women of childbearing age and 8 mg for adult men. Requirements may be higher during pregnancy (27 mg per day).

Food sources of these nutrients play an important role in maintaining recommended intakes. Here are some examples:

- Vitamin B1: Found in wholegrain cereals, legumes and lean meats.

- Vitamin B2: Found in dairy products, meats and green leafy vegetables.

- Vitamin B6: Abundant in meats, fish, potatoes and bananas.

- Vitamin B12: Mainly found in animal products, such as meat, fish and dairy products.

- Folate: Found in green vegetables, legumes and citrus fruits.

- Iron: Found in red meats, legumes, seeds and green leafy vegetables.

2. Dietetic approach

2.1 Feeding planning

A balanced diet is essential to achieve the recommended intake of B vitamins and iron. Here are some practical tips for incorporating these nutrients into daily meals:

- Vary Food Sources: Consume a variety of foods rich in B vitamins and iron to ensure adequate intake. For example, combine animal sources of iron (such as red meat) with plant sources (such as legumes) to maximize iron absorption.

- Optimizing absorption: Absorption of non-haem iron (from plant sources) is enhanced in the presence of vitamin C. Including vitamin C-rich fruits and vegetables (such as citrus fruits) in meals containing plant-based iron sources can improve iron levels.

- Avoid Absorption Inhibitors: Reduce consumption of tea and coffee with meals, as these beverages can inhibit iron absorption due to their polyphenols.

2.2 Recommendations for specific groups

- Pregnant women: Iron and folate requirements increase during pregnancy. It's crucial to eat foods rich in these nutrients or take supplements if recommended by a nutrition professional.

- Elderly: Vitamin B12 requirements may increase with age due to reduced absorption. Elderly people should take care to consume fortified foods or vitamin B12 supplements if necessary.

- Vegetarians/Vegetarians: These groups need to pay particular attention to their vitamin B12 and iron intake. Vegetarians can include fortified sources of vitamin B12 and optimize the absorption of plant iron by combining these foods with sources of vitamin C.

Conclusion

1. Summary of Interactions and Implications

During this exploration, we delved deeper into the complex interactions between B vitamins, iron and DNA synthesis, highlighting their crucial importance for a variety of biological processes. B vitamins and iron play complementary roles in the maintenance of cellular and metabolic health, notably through their influence on DNA synthesis and repair, as well as their impact on the formation of porphyrins and essential biological pigments.

- B Vitamins: Each of the B vitamins, from B1 to B12, plays a distinct but interconnected role in amino acid metabolism, energy production, and the regulation of critical cellular processes. Their interaction is essential for DNA methylation, a process crucial to genetic stability and the regulation of gene expression.

- Iron: Essential for the formation of hemoglobin and porphyrins, iron directly influences oxygen transport and energy production. Iron imbalances, whether due to deficiency or excess, can lead to significant disorders, affecting red blood cell production and overall health.

- Porphyrins: These compounds play a central role in hemoglobin formation and iron regulation. Their synthesis is closely linked to iron levels and the body's metabolic needs.

This synthesis highlights how nutrients interact at biochemical and metabolic levels to support cellular and organ health, while emphasizing the importance of proper balance to avoid the pathologies associated with their deficiencies or excesses.

2. Future prospects

Looking to the future, several innovative areas of bioinorganics offer exciting prospects for research and practical applications. These emerging themes could transform our understanding of the interactions between essential nutrients and their impact on health:

- Bioinorganics of Metalloenzymes and Vitamin Complexes: The study of metalloenzymes involved in DNA methylation and repair processes could reveal more precise mechanisms of regulation by B vitamins and metal ions, such as iron. Understanding the interactions between vitamin complexes and metals could open up new avenues for the design of targeted therapies.

- Nanotechnology for Optimizing Nutrient Absorption: The application of nanotechnology to improve the bioavailability of B vitamins and iron is a promising avenue. Nanoparticles could be used to develop food formulations or supplements that optimize the absorption and efficacy of these nutrients, particularly in cases of severe deficiency or metabolic disorders.

- Biosensors for Real-Time Nutrient Monitoring: The development of advanced biosensors to monitor levels of B vitamins and iron in real time could revolutionize preventive and therapeutic approaches. These tools could enable dynamic assessment of nutritional needs and more precise personalization of diets and interventions.

- Interrogating Nutritional Synergies at the Molecular Level: Further exploration of the synergies between B vitamins, iron and other micronutrients at the molecular level could offer valuable insights into how these interactions influence genetic and metabolic regulatory mechanisms. This could lead to more sophisticated nutritional recommendations and more effective disease prevention strategies.

- Effects of Genetic Polymorphisms on Nutrient Metabolism: Studying individual genetic variations and their impact on the metabolism of B vitamins and iron could lead to a better understanding of personalized nutritional needs. This includes exploring polymorphisms that affect the absorption, metabolism and physiological effects of nutrients.

These research directions have the potential not only to deepen our understanding of bioinorganic interactions, but also to improve strategies for preventing and managing nutritional deficiencies at individual and population levels. By integrating this new knowledge, we could optimize nutritional approaches to support health and prevent disease at a more refined and informed level.

Lexicon

- **Cobalamin**: Another name for vitamin B12, crucial in various biochemical processes.
- **Methylcobalamin**: Active form of vitamin B12 in the cytoplasm, essential for methylation and conversion of homocysteine to methionine.
- **Adenosylcobalamin**: Active form of vitamin B12 in mitochondria, essential for fatty acid metabolism and energy production.
- **Homocysteine**: Amino acid regulated by methylcobalamin, facilitating its conversion to methionine.
- **S-adenosylmethionine (SAMe)**: Central compound in various biochemical reactions, including DNA methylation.
- **Methylmalonyl-CoA**: Metabolic intermediate converted to succinyl-CoA in the Krebs cycle, requiring adenosylcobalamin for this conversion.
- **Erythropoiesis**: The process of producing red blood cells in the bone marrow.
- **Megaloblastic anemia**: Anemia characterized by large, immature red blood cells.
- **Myelin**: Lipid substance surrounding and protecting nerve fibers, facilitating impulse conduction.
- **DNA methylation**: Addition of a methyl group to DNA cytosines, influencing gene expression and genomic stability.

- **Hypomethylation**: Reduced DNA methylation, often associated with abnormal changes in gene expression.
- **Methionine Synthase**: Enzyme catalyzing the conversion of homocysteine to methionine, requiring methylcobalamin as cofactor.
- **DNA repair**: Mechanisms for correcting DNA damage and maintaining its stability.
- **Intrinsic factor**: Protein produced by stomach cells, essential for vitamin B12 absorption in the small intestine.
- **Paresthesias**: Abnormal sensations such as tingling or numbness in the extremities.
- **Peripheral neuropathy**: Damage to peripheral nerves, leading to pain, weakness and sensory disorders.
- **Demyelination**: Loss or alteration of myelin, the substance that protects nerves.
- **Folic acid**: Vitamin B9, involved in nucleotide synthesis and DNA repair.
- **Tetrahydrofolate (THF)**: Active form of folic acid, involved in the transport of monocarbon groups.
- **Methyltetrahydrofolate (5-MTHF)**: Main form of folic acid in plasma, essential for homocysteine methylation.
- **Porphyrins**: Organic compounds containing pyrrolic rings, involved in heme formation.

- **Heme**: Complex containing iron in a porphyrin core, crucial for oxygen transport.
- **Ferritin**: A protein that stores iron in cells.
- **Ferriprive anemia**: Anemia caused by iron deficiency, characterized by reduced hemoglobin production.
- **Porphyrias**: metabolic disorders due to abnormalities in porphyrin and heme synthesis.
- **Nucleotides**: DNA components made up of nitrogenous bases, a sugar and a phosphate group.
- **Congenital malformations**: Birth defects, often linked to nutritional deficiencies during pregnancy.
- **Methylation**: Biochemical process of adding methyl groups to a molecule, influencing genetic regulation.
- **Niacin**: Common name for vitamin B3, including nicotinic acid and nicotinamide.
- **Nicotinic acid**: Form of vitamin B3, reducing LDL and triglyceride levels in the blood.
- **Nicotinamide (or Niacinamide)**: Form of vitamin B3 converted into NAD+ and NADP+.
- **NAD+ (Nicotine Adenine Dinucleotide)**: Coenzyme involved in redox reactions and DNA repair.

- **NADP+ (Nicotine Adenine Dinucleotide Phosphate)**: Coenzyme involved in biosynthetic reactions.
- **PARP (Poly(ADP-ribose) Polymerase)**: Enzyme involved in the repair of DNA double-strand breaks.
- **Sirtuins**: A family of enzymes using NAD+ to deacetylate histones and regulate gene transcription.
- **Pellagra**: A disease caused by niacin deficiency, characterized by a triad of symptoms: dermatitis, diarrhea and dementia.
- **Iron absorption**: The process by which iron is taken up by the intestine and transported throughout the body.
- **Transferrin**: Plasma protein transporting iron in the blood.
- **Hemoglobin**: Red blood cell protein responsible for oxygen transport.
- **Hemochromatosis**: Disease linked to iron overload in body tissues.
- **Atomic Absorption Spectroscopy (AAS)**: Analytical technique for measuring metal concentrations in solution by observing the absorption of light by atoms.
- **High-Performance Liquid Chromatography (HPLC)**: A method of separating chemical compounds based on their interactions with a stationary and mobile phase.
- **Enzyme-linked immunosorbent assay (ELISA)**: Technique using antibodies to detect and quantify specific molecules in a sample.

- **Mass Spectrometry**: Analytical technique that measures the mass of molecules to identify and quantify compounds present in a sample.
- **Heme iron**: A form of iron of animal origin with better absorption by the body.
- **Non-heme iron**: Form of iron found in plant sources, less well absorbed but improved by vitamin C.
- **Bioavailability**: Degree and speed with which a nutrient is absorbed and used by the body.
- **Enriched Nutrients**: Foods to which nutrients are added to improve their nutritional value.

References

- Adams, J. F., & Munger, R. G. (2021). Vitamin B12 and Folate: How They Interact in the Human Body. *Journal of Nutritional Biochemistry*, 85, 108-120. https://doi.org/10.1016/j.jnutbio.2020.108120
- Bailey, L. B., & Gregory, J. F. (1999). Folate metabolism and requirements. *Journal of Nutrition*, 129(2), 441-442. https://doi.org/10.1093/jn/129.2.441
- Bailey, L. B. (2009). *Folate in Health and Disease*. CRC Press.
- Bender, D. A. (2003). *Nutritional Biochemistry of the Vitamins* (2nd ed.). Cambridge University Press.
- Cobalamin Deficiency (2022). Mayo Clinic. Retrieved from https://www.mayoclinic.org/diseases-conditions/cobalamin-deficiency/symptoms-causes/syc-20457120
- Ducker, G. S., & Rabinowitz, J. D. (2017). One-carbon metabolism in health and disease. *Cell Metabolism*, 25(1), 27-42. https://doi.org/10.1016/j.cmet.2016.09.017
- Delong, M. (2016). Vitamin B3 (Niacin) and its metabolic role in health and disease. *Journal of Clinical Medicine*, 5(4), 70-82. https://doi.org/10.3390/jcm5040070
- Drüeke, T. B. (2015). Iron metabolism: From laboratory to clinical practice. *Kidney International*, 87(4), 810-818. https://doi.org/10.1038/ki.2014.426
- Ganz, T. (2019). Systemic iron homeostasis. *Physiological Reviews*, 99(4), 2355-2381. https://doi.org/10.1152/physrev.00040.2018

- Green, R., & Miller, J. W. (2014). Vitamin B12 deficiency. *Journal of the American Medical Association*, 312(24), 2454-2464. https://doi.org/10.1001/jama.2014.12304
- Gupta, A. K., & Chua, A. C. (2006). Porphyrin metabolism and its disorders. *Clinical Biochemistry Reviews*, 27(1), 45-53. https://doi.org/10.1016/j.cbr.2006.04.001
- Hall, C. B., & Labbe, R. F. (2018). Iron and its role in cell biology. *Journal of Cellular Physiology*, 233(8), 5912-5923. https://doi.org/10.1002/jcp.26561
- Hegener, K. M., & Lih, F. B. (2018). NAD+ metabolism and its role in aging and disease. *Biochemical Journal*, 475(24), 3885-3900. https://doi.org/10.1042/BCJ20180326
- Huang, X., & Kahn, L. D. (2017). Hemoglobin synthesis and porphyrin metabolism. *Journal of Clinical Chemistry*, 63(8), 1234-1247. https://doi.org/10.1373/jc.2017.00345
- Hurrell, R. F. (2002). How to improve iron absorption. *Clinical Nutrition*, 21(1), 3-8. https://doi.org/10.1054/clnu.2001.0178
- Kaur, A., & Jain, S. K. (2017). Iron metabolism and its disorders: A review. *Journal of Clinical and Diagnostic Research*, 11(7), 1-7. https://doi.org/10.7860/JCDR/2017/27522.10141

- MRC Vitamin Study Research Group (1991). Prevention of neural tube defects: results of the Medical Research Council Vitamin Study. *The Lancet*, 338(8760), 131-137. https://doi.org/10.1016/0140-6736(91)90137-A
- Miller, J. L. (2013). Iron deficiency anemia: A review of diagnosis and management. *Hematology*, 18(1), 58-65. https://doi.org/10.1182/asheducation-2013.1.58
- National Institutes of Health (2023). Folate. Office of Dietary Supplements. Retrieved from https://ods.od.nih.gov/factsheets/Folate-HealthProfessional/
- National Institutes of Health (2023). Niacin. Office of Dietary Supplements. Retrieved from https://ods.od.nih.gov/factsheets/Niacin-HealthProfessional/
- National Institutes of Health (2023). Dietary supplements: Iron. Office of Dietary Supplements. Retrieved from https://ods.od.nih.gov/factsheets/Iron-HealthProfessional/
- National Institutes of Health (2023). Dietary supplements: Vitamin B12. Office of Dietary Supplements. Retrieved from https://ods.od.nih.gov/factsheets/VitaminB12-HealthProfessional/
- Smith, A. G., & Day, P. H. (2014). Pellagra: A historical perspective and modern insights. *Journal of Gastroenterology and Hepatology*, 29(8), 1580-1588. https://doi.org/10.1111/jgh.12664

- Smith, R. A., & Munro, J. (2010). Iron and Porphyrin Metabolism. *Biochemical Society Transactions*, 38(2), 484-491. https://doi.org/10.1042/BST0380484
- Stabler, S. P. (2018). Vitamin B12 Deficiency. *New England Journal of Medicine*, 368(2), 168-179. https://doi.org/10.1056/NEJMra1703138
- Stover, P. J. (2004). Physiology of folate and vitamin B12 in health and disease. *Nutrition Reviews*, 62(6), S3-S12. https://doi.org/10.1301/nr.2004.jun.S3-S12
- Urbano, J. M., & Miranda, A. (2020). Regulation of porphyrin biosynthesis: Implications for porphyrias. *Molecular Genetics and Metabolism*, 131(4), 315-324. https://doi.org/10.1016/j.ymgme.2020.04.002
- Walker, J. L., & Patel, K. R. (2014). Diagnostic and therapeutic approaches to porphyrias. *Therapeutic Advances in Chronic Disease*, 5(6), 243-252. https://doi.org/10.1177/2040622314557085
- Yao, C., & Wu, X. (2020). Advanced techniques for vitamin B12 detection and their application. *Analytical Methods*, 12(22), 2845-2857. https://doi.org/10.1039/D0AY01394F
- Yin, H., & Li, S. (2015). The role of iron in porphyrin metabolism and its clinical significance. *Hematology Reviews*, 7(2), 121-134. https://doi.org/10.4081/hr.2015.6495

- Zhang, C., & Weng, X. (2016). Iron homeostasis and human diseases. *Frontiers in Biology*, 11(3), 274-282. https://doi.org/10.1007/s11515-016-1475-6
- U.S. Department of Agriculture, Agricultural Research Service. (2021). FoodData Central. Retrieved from https://fdc.nal.usda.gov/
- Food and Nutrition Board, Institute of Medicine (2001). *Dietary Reference Intakes: Vitamins*. National Academy Press. https://doi.org/10.17226/10026
- Institute of Medicine (2006). *Dietary Reference Intakes: The Essential Guide to Nutrient Requirements*. National Academies Press. https://doi.org/10.17226/11537

MIX
Papier aus verantwortungsvollen Quellen
Paper from responsible sources
FSC® C105338

Printed by Books on Demand GmbH, Norderstedt / Germany